Little Armored Ones

The Armadillo Kingdom, Illustrated

Ryan E. Felton

Design & layout by Rachel Leigh

Little Armored Ones:
The Armadillo Kingdom, Illustrated.
Copyright 2018 Ryan E. Felton
All rights reserved.
ISBN-13: 978-1984055101
ISBN-10: 1984055100

There is some bias in the following pages. The author adores armadillos—every kind (but especially nine-banded armadillos).

There are at least 20 types of armadillo in the world. Maybe you've seen one at a zoo or on the side of the road. But most of them, you've probably never heard of. A few of them might surprise you, like the hamster-sized pink fairy armadillo. Or the very vocal screaming hairy armadillo. Or the Muppet-looking hairy long-nosed armadillo.

They all play a role in their ecosystems. They build homes for other animals, keep termite and other destructive insect populations in check, and even gave us the cure for leprosy.

(Yes, certain armadillos *can* carry leprosy—in the same way we have that potential, thanks to our body temperature. Assuming every armadillo has the disease is a little like assuming every human does. But don't bug wild animals regardless, maybe?)

Guacamole fans, take a silent moment to thank the noble armadillo: Prehistoric 'dillos (called Glyptodons) were critical to the spread and growth of avocados.

I hope this book inspires armadillo appreciation in places that don't get much chance to see or learn about them. As a native Hoosier, I've had to do a lot of my own digging to build up my lifelong passion. I hope this serves as a handy reference and inspires you to learn more—about all kinds of unusual, unique animals.

Ryan E. Felton

Nine-Banded Armadillo 7

Seven-Banded Armadillo 9

LONG-NOSED

Llanos Long-Nosed Armadillo 11

Southern Long-Nosed Armadillo 13

Greater Long-Nosed Armadillo 15

Hairy Long-Nosed Armadillo 17

Yepes's Mulita 19

LONG-NOSED

This book is approximately 16 inches wide. The nine-banded armadillo measures 18 inches from nose to rump.

Nine-Banded Armadillo

a.k.a. Panzerschwein
a.k.a. "Hoover Hog"

Dasypus novemcinctus

Close your eyes. Think, "armadillo." I'll bet you're picturing the nine-banded armadillo.

As the only species found in the U.S.—and God's perfect creature—this critter is the Texas state mascot. She mostly eats insects and gets around with her incredible sense of smell. She can sniff out a grub buried six inches deep. She's pretty much blind, though.

Nine-banded armadillos always give birth to four identical pups of the same gender. (I think this would make a fantastic premise for a sitcom. Chuck Lorre has not returned my calls.) A nine-banded armadillo can jump four feet high, hold her breath for six minutes, and run up to thirty miles an hour. I'm not kidding. In southern states, people train and race them.

I might as well tell you: Some people eat these little angels. In the 1930s, they were called "Hoover hogs" because people blamed Herbert Hoover for the Depression that stuck them with armadillo chops for dinner.

Nine-banded armadillos have a bigger fan club than just myself. A bird called the fan-tailed warbler follows our armored pal around to pick up the worms she rustles up when she digs. What a saint.

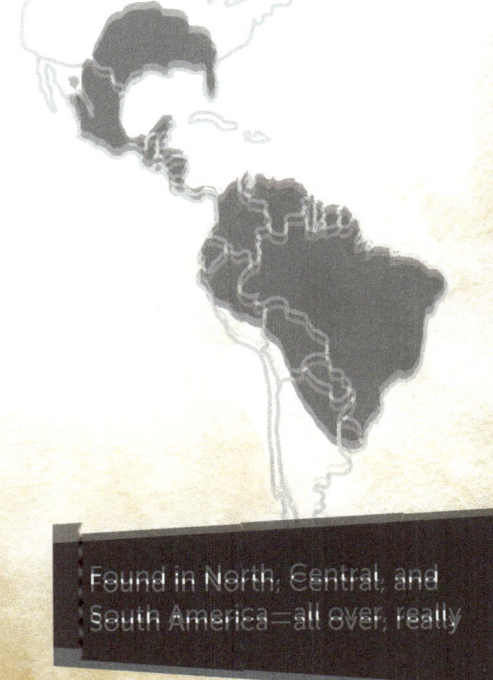

Found in North, Central, and South America—all over, really

About as big as your cat, maybe?

LONG-NOSED

15 inches: a fancy MacBook Pro.

Seven-Banded Armadillo

a.k.a. Brazilian Lesser Long-Nosed Armadillo

Dasypus septemcinctus

Seven-banded armadillos prefer their own company and don't mingle much.

Occasionally, though, they'll share a burrow with other seven-banded armadillos of the same sex—essentially forming a seven-banded frat house. Speaking of fraternity, they can identify their siblings by smell. It's like when the aroma of English Leather is in the air and your mom's like, "Oh, is your Great-Uncle Mort here?"

Found in South America

LONG-NOSED

12 inches or so, without the tail.

Llanos Long-Nosed Armadillo

Dasypus sabanicola

This armadillo is accustomed to the heavy flooding typical of its home region.

Imagine a nine-banded armadillo being cast in the Costner role of *Waterworld*, and you more or less have a Llanos long-nosed.

Found in the Llanos grasslands, Colombia and Venezuela
Conservation status: NEAR-THREATENED

LONG-NOSED

12 inches, head-to-hind.

Southern Long-Nosed Armadillo

Dasypus hybridus

The long-nosed chaps of the South build nests in their burrows and hide the entrances with leaves and rocks.

Our buddy here is either pretty focused or really blind: When sniffing out food, he's been known to bump right into humans.

He has glands on his eyelids, feet, and ears that secrete a yellow liquid (not pee, okay?) to attract a mate.

Found in Brazil, Argentina, Uruguay
Conservation status: **NEAR-THREATENED**

LONG-NOSED

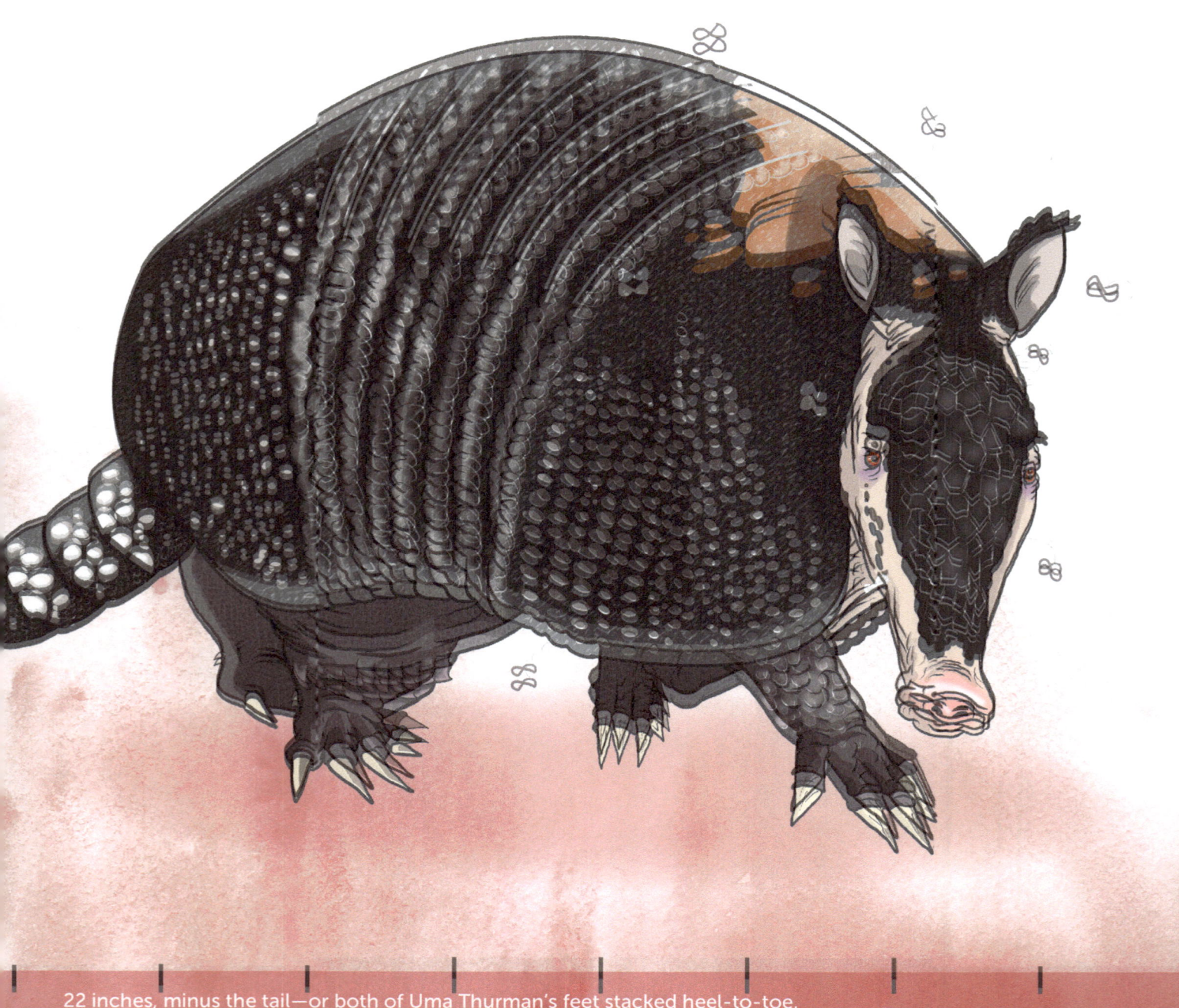

22 inches, minus the tail—or both of Uma Thurman's feet stacked heel-to-toe.

Greater Long-Nosed Armadillo

Dasypus kappleri

Do not look the largest of the long-nosed armadillos directly in the eye:

His kind dwells in muggy swamplands, only coming out at night. He's got spurs on his back heels, projecting scales on his knees, and extra armor on the back of his hind legs. He is ready to fight you.

Most of what we know about this species, we learned from the Matsés tribe of the Amazon. Researchers kind of took the locals at their word and called it a day.

He is often accompanied by a cloud of small white flies. Few armadillos are vocal, but the greater long-nosed armadillo will growl—or almost bark—when angry.

He is always angry.

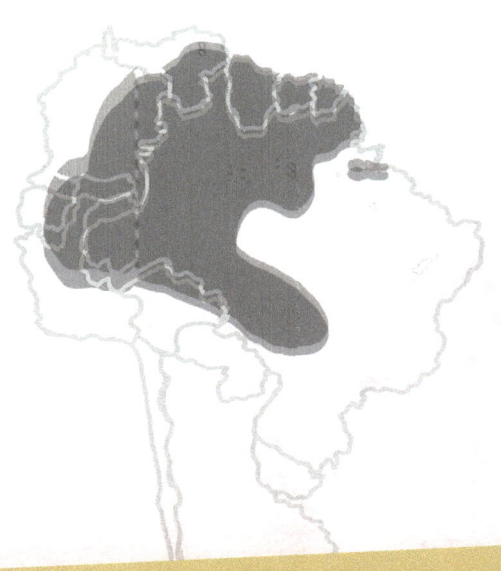

Found in the Brazilian Amazon basin, plus Ecuador, Bolivia, and Peru

LONG-NOSED

13 inches, sans tail—or a generous Subway sandwich.

Hairy Long-Nosed Armadillo
a.k.a. Wooly Armadillo

Dasypus pilosus

One of the least well-known armadillos (somehow), the hairy long-nosed armadillo lives in the tropical habitat of the Andes Mountains. I can't imagine what the humidity does to his hairdo. All that fuzz grows through tiny pores in his carapace.

He's a point of contention among the scientific community. Some researchers think he's so unique he should be classified to a different genus. He's definitely got his own thing going on.

Found in Peru. Conservation status: **DATA-DEFICIENT, LIKELY VULNERABLE**

LONG-NOSED

Size unknown.

a.k.a. Yunga's Lesser Long-nosed Armadillo

Yepes's Mulita

Dasypus yepesi

Shrouded in mystery, this clandestine creature is named for the researcher Jose Yepes. Good luck finding out anything else about it. Seriously: Google it. You'll come up empty every time. It is the J.D. Salinger of armadillos.

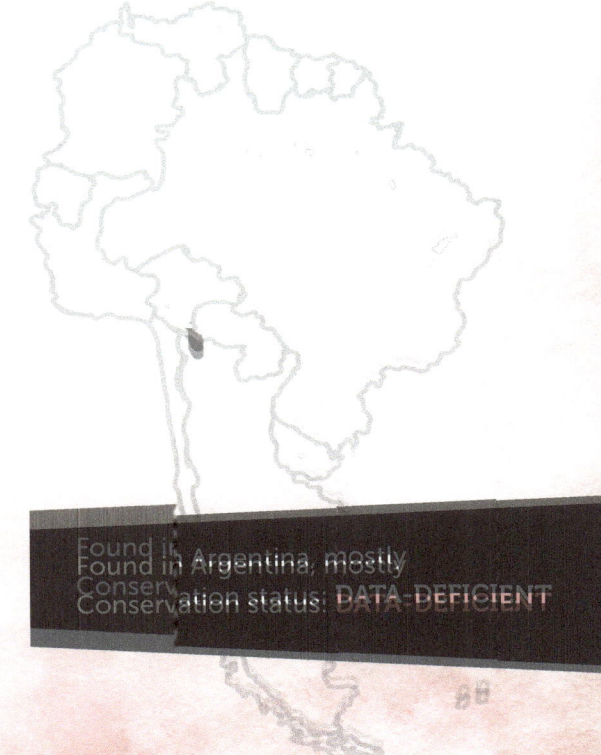

Found in Argentina, mostly
Conservation status: DATA-DEFICIENT

Screaming Hairy Armadillo 23

Andean Hairy Armadillo 25

HAIRY

Big Hairy Armadillo 27

HAIRY

About 10 inches (not including tail)

Screaming Hairy Armadillo

a.k.a. Dwarf Hairy Armadillo

a.k.a. Crying Armadillo

a.k.a. Small Hairy Armadillo

Chaetophractus vellerosus

Found in Gran Chaco and Pampas regions, South America

Boasting not one but two adjectives, the screaming hairy armadillo is perhaps the most aptly named of all: When threatened, he emits a shrill cry. Sometimes he emits a shrill cry when not threatened, simply to test the patience of his neighbors.

It is said that half the volume of his stomach might be occupied by sand accidentally ingested while snarfing plants and bugs. So I'm wondering if the screaming might be attributed to horrid indigestion.

Argentinians have a thing about using screaming hairy armadillo shells to make *charangos*—a kind of stringed musical instrument. I think we have established that he has plenty of reasons to scream.

HAIRY

11 inches (not including tail): a sheet of letter-sized paper.

Andean Hairy Armadillo

Chaetophractus ~~nationi~~ vellerosus

Get offa' his property!

The mountain-dwelling Andean hairy armadillo just got Plutoed: While I was writing this book, scientists finally came to an agreement that he doesn't exist. This is the same species as *Chaetophractus vellerosus* (see p.24–25). The only difference is he was raised in a more remote and off-the-map region. Think Nell. He's like Nell.

This grizzled recluse has eighteen bands. Like his mouthy brother, the few teeth he does have lack enamel and never stop growing.

Found in the Puna region, South America
Conservation status: **VULNERABLE**

HAIRY

About 14 inches (tail and batteries not included).

Big Hairy Armadillo
a.k.a. Large Hairy Armadillo

Chaetophractus villosus

The big hairy armadillo is the most common species in South America.

She spends so much time underground, she's got special nasal membranes to get oxygen from soil particles.

She keeps one permanent, carefully and deeply dug burrow, and several shallow temporary residences. For quick escapes and, presumably, wicked ragers.

Found in Argentina, Paraguay

THREE-BANDED

Brazilian Three-Banded Armadillo 31

Southern Three-Banded Armadillo 33

THREE-BANDED

9 inches. (Think Nerf ball.)

Brazilian Three-Banded Armadillo a.k.a. Tatu-bola ("ball armadillo")

Tolypeutes tricinctus

The Brazilian three-banded armadillo drags its nose along the ground as it walks, sniffing out ants and termites.

His armor's a little looser than other armadillos', so he's got more freedom of movement—and his body temperature's regulated a little better. That means he can survive in hotter climes than a lot of his ilk.

Along with its close relative, the Southern three-banded armadillo, this guy can roll into a tight ball for protection. Maybe that's why he had a brush with fame as the 2014 FIFA World Cup mascot.

This species was briefly thought extinct, and is particularly vulnerable. In Brazil, an action plan is in place to protect it.

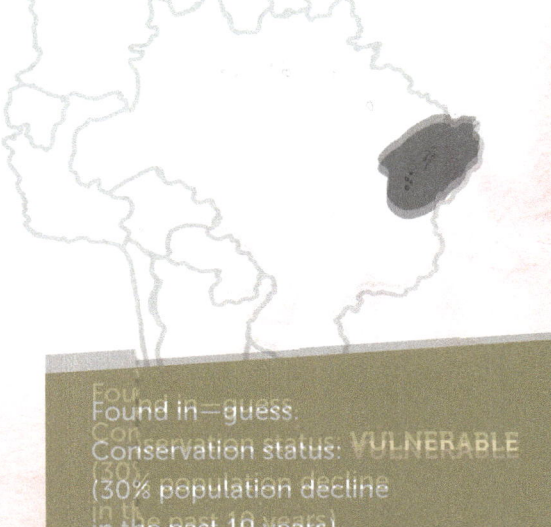

Found in—guess.
Conservation status: **VULNERABLE**
(30% population decline in the past 10 years)

THREE-BANDED

9 inches. (Wike a widdle bunny.)

Southern Three-Banded Armadillo

a.k.a. La Plata Three-Banded Armadillo

Tolypeutes matacus

Along with the Brazilian three-banded armadillo, the Southern three-banded is the only armadillo capable of rolling up into a tight ball—despite popular belief.

This little cutie is not **fossorial** (as in, it does not burrow and live underground). That's notable, considering we're talking about an armadillo here. Hey, guess what? This thing's shell is coated in keratin, which is what your fingernails are made of. You're practically siblings.

Found in South America, especially Argentina
Conservation status: **NEAR-THREATENED**

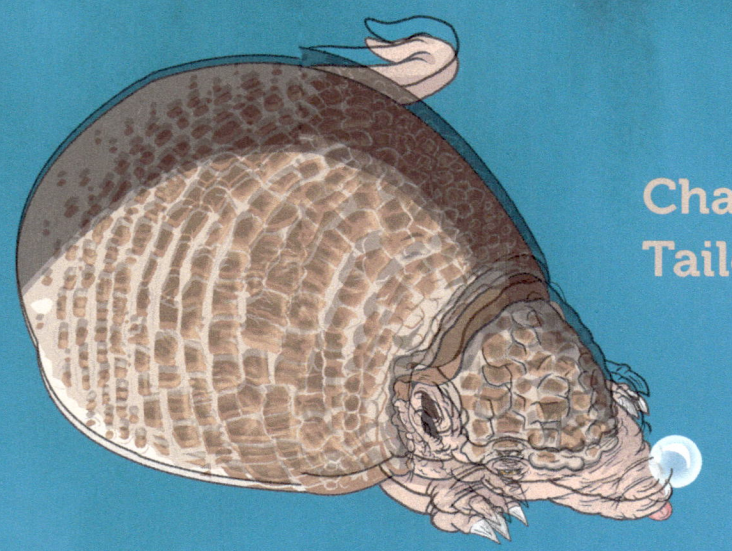

Chacoan Naked-Tailed Armadillo 37

NAKED-TAILED

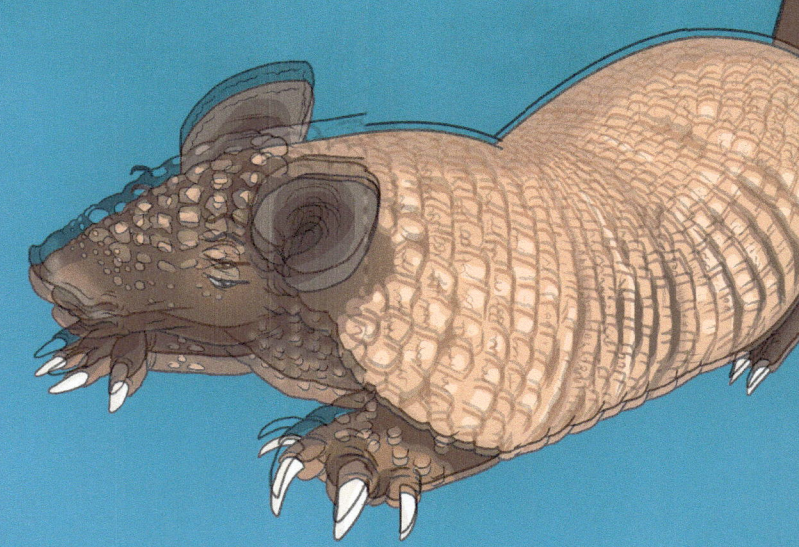

Greater Naked-Tailed Armadillo 39

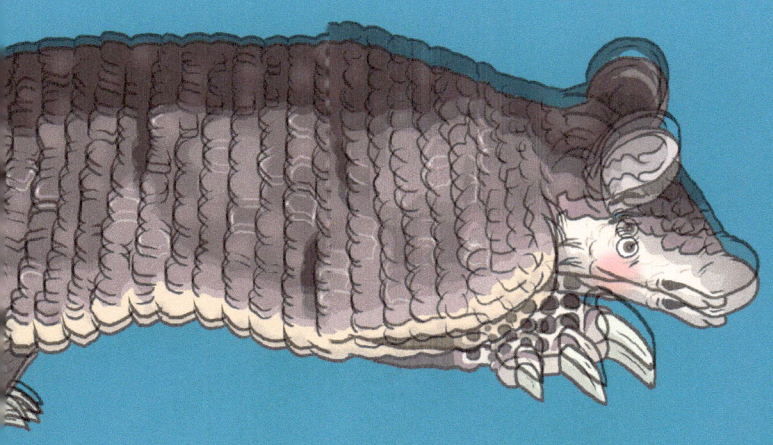

Northern Naked-Tailed Armadillo 41

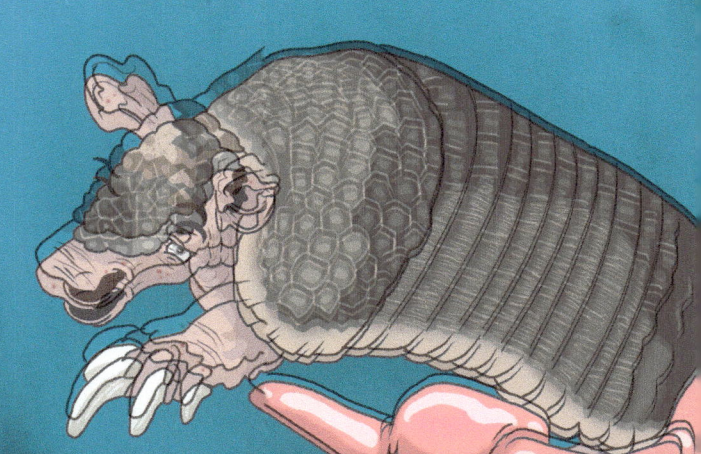

Southern Naked-Tailed Armadillo 43

NAKED-TAILED

15 inches, not including tail. That's the length of a bowling pin.

Chacoan Naked-Tailed Armadillo

Cabassous chacoensis

Here we have a Chacoan naked-tailed armadillo.

Take a look at her unique, fleshy ears. Check out the long, sickle-like middle claws she's sporting. She'd like you to know that, like many armadillo species, she prefers a **xeric** habitat—dry and hot.

Yes, sir. There she is.

Found in the Gran Chaco region, South America
Conservation status: **NEAR-THREATENED**

NAKED-TAILED

18 inches, not including tail. She could hide amongst your American Girl dolls like E.T.

Greater Naked-Tailed Armadillo

a.k.a. Tatouay

Cabassous tatouay

The greater naked-tailed armadillo is a wily architect, building her burrow to face away from prevailing winds. Her species is prevalent in Brazil, so book your flight now if you'd like to see one. But be prepared to get dirty: she hangs out underground mostly.

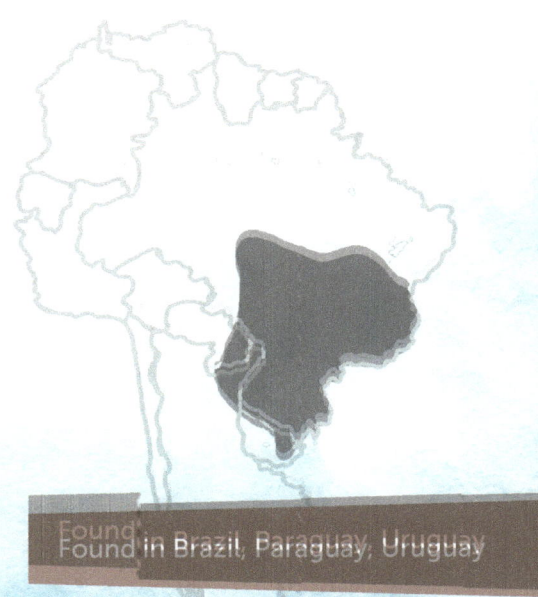

Found in Brazil, Paraguay, Uruguay

NAKED-TAILED

12 inches without the tail.

Northern Naked-Tailed Armadillo

Cabassous centralis

The northern naked-tailed armadillo has some unique physical traits.
First, it has no scales on its ears. Second, the bands of its armor are indistinct. Hair on his belly and tail are sparse to nonexistent. He is said to have a particular and not-pleasant odor. When he digs, he rotates his body like an auger drill—a behavior unique to his kind.

In this species, only one pup is born at a time. An only child, with body odor, fated to never grow chest hair: Puberty was probably not great for him.

Found in Mexico, Colombia, Venezuela, Ecuador
Conservation status: **DATA-DEFICIENT**

NAKED-TAILED

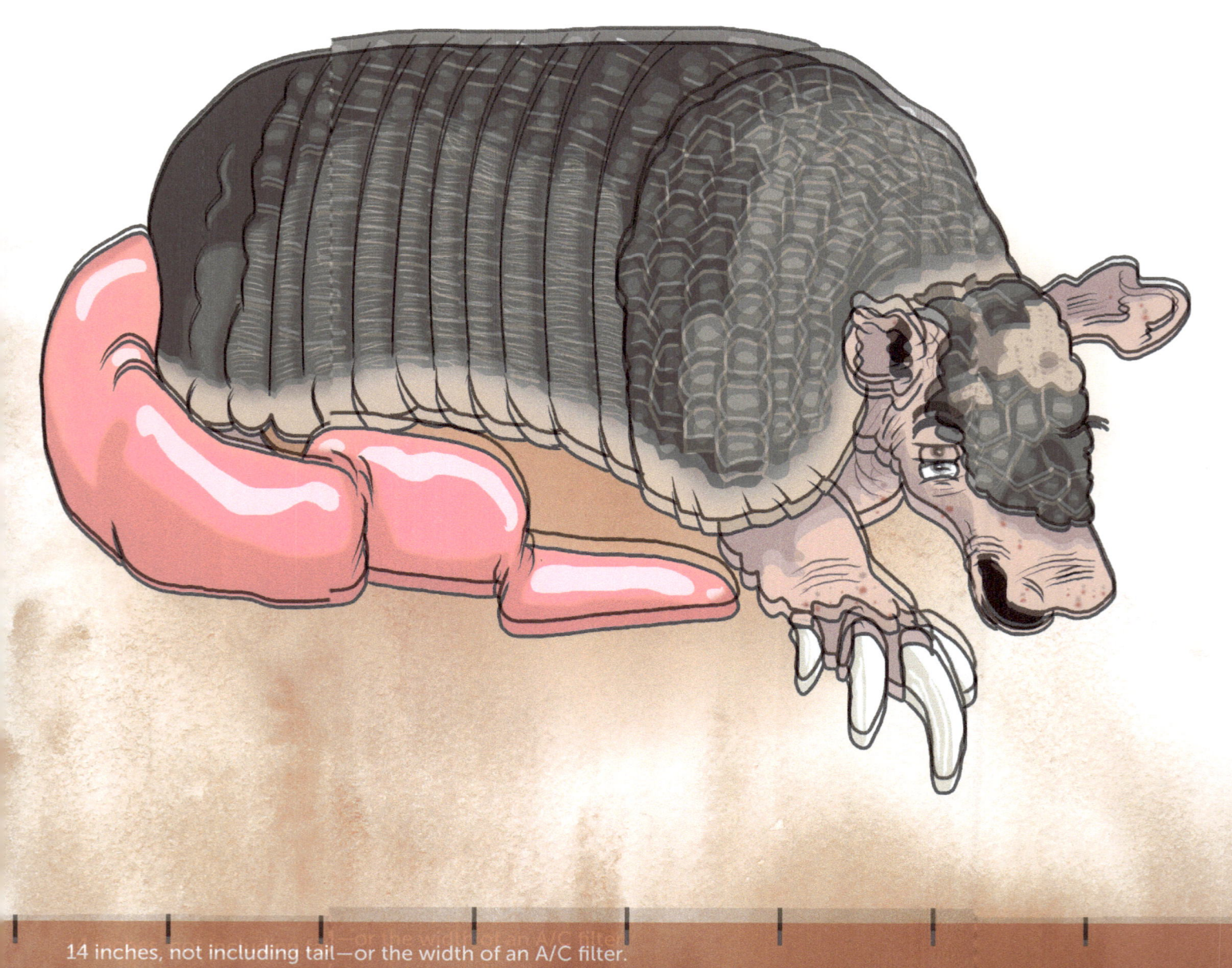

14 inches, not including tail—or the width of an A/C filter.

Southern Naked-Tailed Armadillo

Cabassous unicinctus

Most armadillos have *scutes* covering their tails, but not this guy.

His underbelly and tail are also practically hairless.

He builds his burrows in and around termite nests, so those insects are both his neighbors and his dinner.

Found in many regions of South America

Pichi 47

...AND THE REST

Pink Fairy Armadillo 49

Chacoan Fairy Armadillo 51

Six-Banded Armadillo 53

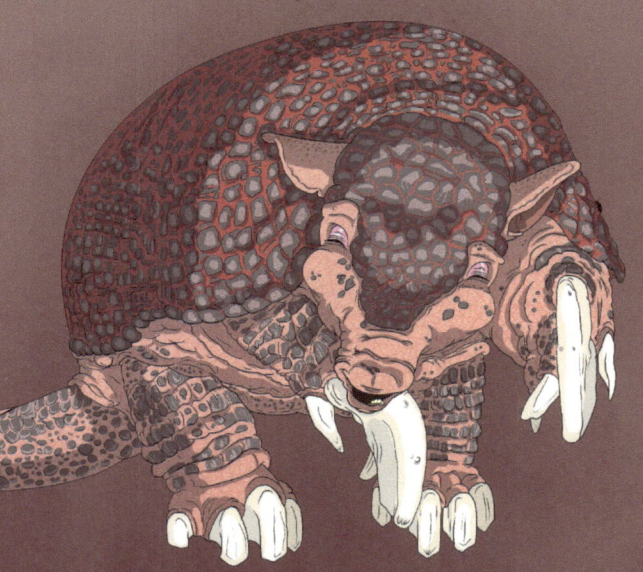

Giant Armadillo 55

AND THE REST

11 inches, not including tail.

Pichi
a.k.a. Dwarf Armadillo
Zaedyus pichiy

The pichi is the only armadillo that hibernates.

She's also diurnal, unlike most species. Notably, she has an extra band of armor around the nape of her neck that other armadillos don't.

Population is declining because they're hunted for sport—and an outbreak of an unknown disease that only affects pichis.

Found in Argentina, Chile
Conservation status: NEAR-THREATENED

AND THE REST

4 inches (hamster-like).

Pink Fairy Armadillo
Chlamyphorus truncatus

a.k.a. Pichiego
a.k.a. "Sand-Swimmer"

Admit it. You just fell in love.

The pink fairy is the smallest of all armadillos; you could hold him in your palm, if you could find one. (You couldn't. In Pokémon terms, they're rarer than Mew.)

His shell is totally separate from his body, unique among all other armadillos. (It's connected to him by a thin membrane along his itty-bitty spine.) HE CAN CHANGE COLOR. His shell is so thin, drastic changes in temperature can alter his pinkish hue to red or white.

He uses his flat butt and spatula-shaped tail to compress the dirt left in his wake as he burrows, so the tunnel won't cave in. The lucky few to have witnessed one of his kind in the wild described them as "sand-swimmers," because they "dig through dirt like fish swim through water."

Now you want one as a pet. I get it. But forget it. Seriously. Pink fairies die in captivity after only a few days.

Found in Argentina
Conservation status: **DATA-DEFICIENT**

AND THE REST

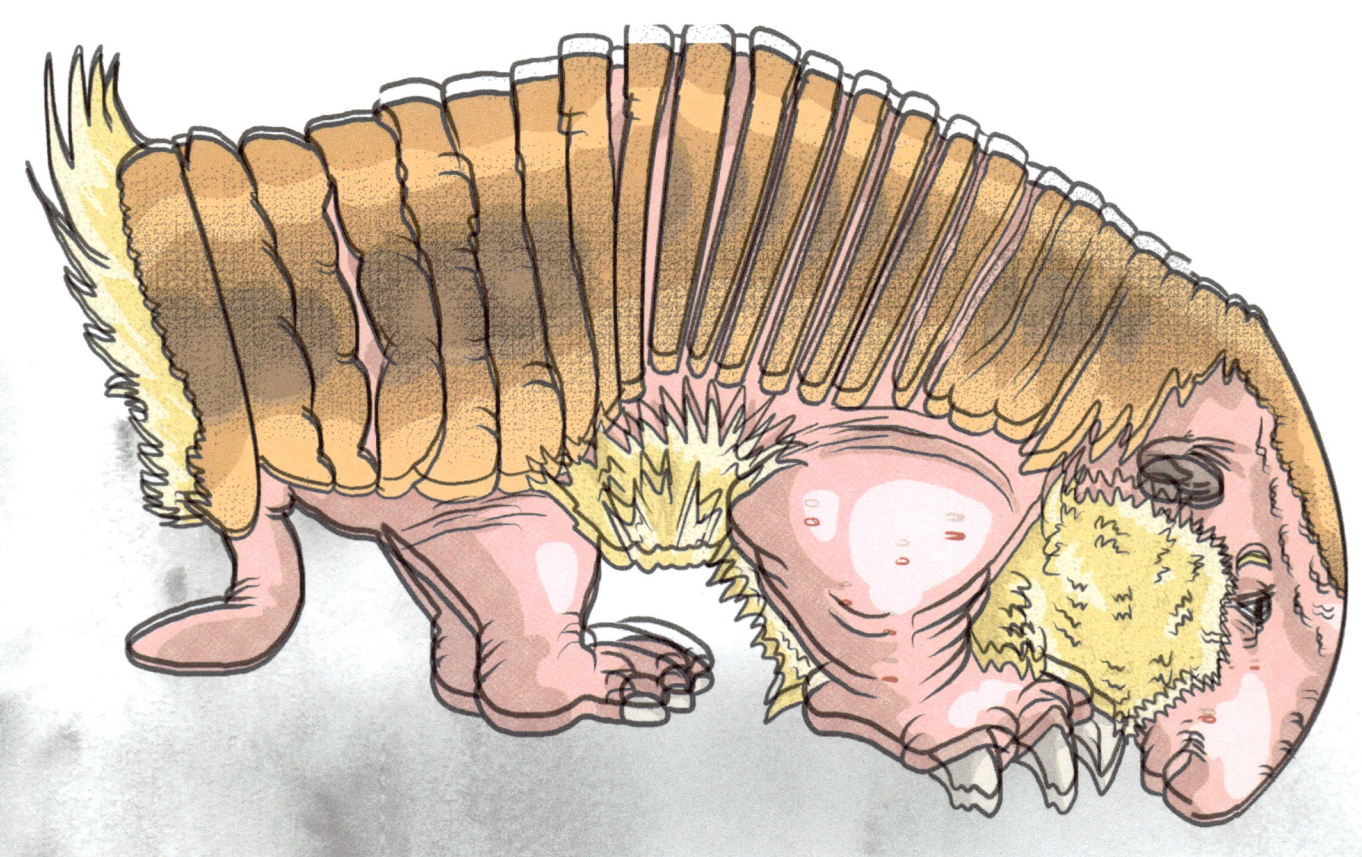

5 inches—would fit nicely in your dollhouse breakfast nook.

Chacoan Fairy Armadillo

a.k.a. Greater Fairy Armadillo
a.k.a. Burmeister's Armadillo

Calyptophractus retusus

Rarely seen, this gnome-like critter lives up to his appearance by hanging out underground.

He uses his disc-like patootie to plug up his shallow burrow. You would too: In southern Bolivia, the Chacoan Fairy Armadillo is killed on sight by Izoceño indigenous people, who believe he is a sign of bad luck.

Baby Chacoan fairies are **precocial**, meaning they're pretty much ready to go the moment they're born. Pink and blind and shrively are about as good as it gets for these things.

Found in South America (Gran Chaco region)
Conservation status: NEAR-THREATENED

AND THE REST

About 20 inches, not including tail.

Six-Banded Armadillo

Euphractus sexcinctus

a.k.a. Yellow Armadillo

a.k.a. Tatu Peludo ("furry armadillo")

The third-largest type of armadillo, the six-banded armadillo's less of a night owl than her kin.

She takes care of business in the daylight. She can store body fat to survive during times of food scarcity.

Does she intimidate you? Because she's actually an aggressively protective mother. She just wants what's best for her kids, okay?

She's pals with the southern naked-tailed armadillo; they share and swap burrows. It's a thing they have.

Lucky for her, the people of her native land aren't interested in eating six-banded armadillos. That's because of a local superstition that her species eats the flesh of human corpses. As far as I know, that's just hearsay. I mean, look at that face.

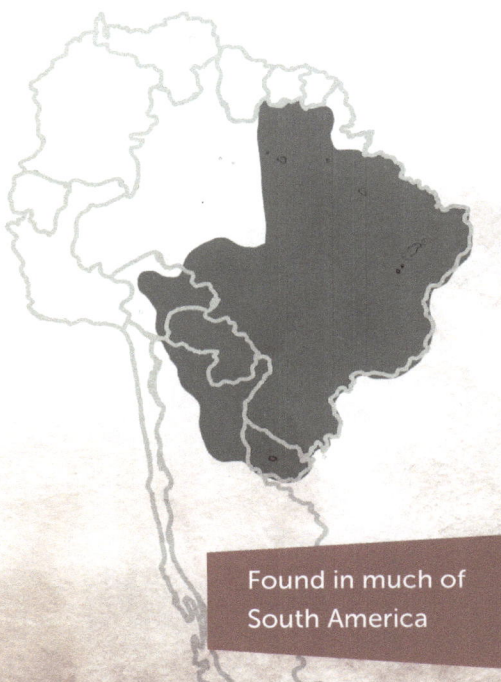

Found in much of South America

AND THE REST

3–4 feet. A toddler could comfortably ride one (comfortable for the toddler, not the armadillo).

Giant Armadillo
Priodontes maximus

a.k.a: Ocarro
a.k.a. Tatu-Canastra

And now the main event: The big one. The heavyweight. The world's largest. Here comes the giant armadillo.

This big boy can eat an entire termite colony. In fact, while most armadillos have only a few back teeth to mash up bugs, the giant armadillo has 100 teeth—more than any other terrestrial mammal. (They're all flat and lack enamel, in true armadillo fashion.) They also have the largest claws, proportionately, of any mammal.

You're looking at the real estate king of the Pantanal. Dozens of species have giant armadillos to thank for their homes. Since he's so nomadic, this armadillo will dig up to 150 burrows a year. When he leaves, lizards, ocelots, tamanduas, and many more snap up the hot property and move in. This makes the giant armadillo crucial to the survival of his entire ecosystem—and he's endangered. Heavy stuff.

Found in the Pantanal region, and scattered across South America
Conservation status: **VULNERABLE**

Glossary

You might want to know these words.

Carapace
A fancy word for the armadillo's shell or armor

Chlamyphoridae
One of two living families of the armadillo order, Cingulata

Dasypodidae
The other family of the armadillo order.

Diurnal
The opposite of nocturnal; active during the daytime.

Glyptodon
A prehistoric armadillo ancestor

Fossorial Living underground

Lister A male armadillo

Precocial
Self-feeding and independent from the moment of birth

Pup A baby armadillo

Scute
A bony plate, sort of like a big scale; the armadillo's armor is composed of these

Xenarthran
A member of the superorder of mammals that includes armadillos, anteaters, and sloths

Xeric A dry and hot environment

Zed A female armadillo

The text in this book book is indebted to—though produced independently from—the work of Mariella Superina, William James Loughry, and the **IUCN/SSC Anteater, Sloth, and Armadillo Specialist Group**. One may donate to research and conservation efforts at **xenarthrans.org**.

www.ingramcontent.com/pod-product-compliance
Lightning Source LLC
Chambersburg PA
CBHW051922210526
45473CB00006B/2100